OUTRAGEOUS ANIMAL ODDITIES

BY VIRGINIA LOH-HAGAN

 45TH PARALLEL PRESS

Published in the United States of America by Cherry Lake Publishing Group
Ann Arbor, Michigan
www.cherrylakepublishing.com

Reading Adviser: Beth Walker Gambro, MS Ed., Reading Consultant, Yorkville, IL
Book Designer: Melinda Millward

Photo Credits: Cover: © Neil Bromhall/Shutterstock; Page 1: © Neil Bromhall/Shutterstock; Page 5: © NaturePhoto/Shutterstock; Page 6: © Sergey Uryadnikov/Shutterstock; Page 7: © Seatraveler/Dreamstime; Page 8: © Ryan M. Bolton/Shutterstock; Page 10: © Andrea Izzotti/Adobe Stock, © feathercollector/Shutterstock; Page 11: © Agami Photo Agency/Shutterstock; Page 12: © Matt Cornish/Shutterstock, © Jennifer Brady/Dreamstime; Page 13: © Kensho Photographic/Shutterstock; Page 14: © javarman/Shutterstock; Page 15: © dennisvdw/Thinkstock; Page 16: © Ash/Adobe Stock, © Jacqui Martin/Shutterstock; Page 17: © Lukas/Adobe Stock; Page 18: © Lukas/Adobe Stock; Page 20: © Kasparart/Dreamstime; © NOAA/NMFS/SEFSC Pascagoula Laboratory; Collection of Brandi Noble, NOAA/NMFS/SEFSC/http://www.flickr.com/ CC-BY-2.0; Page 21: © Neil Bromhall/Shutterstock, © panparinda/Shutterstock

Graphic Element Credits: Cover, multiple interior pages: © paprika/Shutterstock, © Silhouette Lover/Shutterstock, © Daria Rosen/Shutterstock, © Wi_Stock/Shutterstock

Copyright © 2023 by Cherry Lake Publishing Group
All rights reserved. No part of this book may be reproduced or utilized in any form or by any means without written permission from the publisher.
45TH Parallel Press is an imprint of Cherry Lake Publishing Group.

Library of Congress Cataloging-in-Publication Data

Names: Loh-Hagan, Virginia, author.
Title: Outrageous oddities / by Virginia Loh-Hagan.
Description: Ann Arbor, Michigan : Cherry Lake Publishing, [2023] | Series: Wild Wicked Wonderful Express. | Audience: Grades 2-3 | Summary: "Which animals are outrageously odd? This book explores the wild, wicked, and wonderful world of the strangest animals in the animal kingdom. Series is developed to aid struggling and reluctant young readers with engaging high-interest content, considerate text, and clear visuals. Includes table of contents, glossary with simplified pronunciations, index, sidebars, and author biographies"—Provided by publisher.
Identifiers: LCCN 2022042726 | ISBN 9781668919712 (hardcover) | ISBN 9781668920732 (paperback) | ISBN 9781668923399 (pdf) | ISBN 9781668922064 (ebook)
Subjects: LCSH: Animals—Adaptation—Juvenile literature.
Classification: LCC QH546 .L64 2023 | DDC 591.4—dc23/eng/20220914
LC record available at httA://lccn.loc.gov/2022042726

Cherry Lake Publishing Group would like to acknowledge the work of the Partnership for 21st Century Learning, a Network of Battelle for Kids. Please visit www.battelleforkids.org/networks/p21 for more information.

Printed in the United States of America

About the Author
Dr. Virginia Loh-Hagan is an author, university professor, former classroom teacher, and curriculum designer. She loves all the odd things about her friends. Odd is special! Odd is cool! She lives in San Diego with her tall husband and very naughty dogs.

Table of Contents

Introduction 4
Proboscis Monkeys 6
Flying Fish 10
Cassowaries 12
Aye-Ayes 14
Platypuses 16
Anglerfish 20

Consider This! 24
Glossary 24
Index 24

Introduction

Animals are odd. They look odd. They do odd things. They stand out. They're special.

They're odd for different reasons. Animals develop ways to **survive**. Survive means to stay alive. They're built for their **environment**. Environment means their home. They **adapt** to where they live. Adapt means change.

Some animals are extreme **oddities**. Oddities are strange things. Their looks are odder than most. Their **habits** are odder than most. Habits are ways of doing things. They're the most exciting oddities in the animal world!

Over time, animals have evolved, or changed, to survive.

Proboscis Monkeys

Proboscis monkeys live in Asia. A proboscis is a long, flexible nose. These monkeys live in groups. Each group has 1 male. The group also has several females and their young. Males can be 50 pounds (23 kilograms). Females are half that size.

Males have big noses. Their noses hang low. They use their noses to attract females. Male noses make sounds. They honk. The honks scare off other males.

Male proboscis monkeys' noses slowly grow until adulthood.

Proboscis monkeys are among the largest of Asia's monkeys.

Proboscis monkeys eat leaves, seeds, and fruits. They have an odd eating system. They let food get soft in their stomachs. They throw up large bits. Then they chew it again. Chewing breaks down food.

When Animals Attack!

People are scared of the "Vampire Beast of North Carolina." This animal is blamed for killing animals. It mainly killed dogs. It crushed their heads. It tore their bodies. It drained blood from their dead bodies. This animal attacked in 1953, 2003, and again in 2013. It seems to come and go. People say it has the body of a bear and the head of a cat. Experts say it doesn't exist. People are probably seeing a wildcat.

Flying Fish

Flying fish live in the Atlantic Ocean. They're **prey**. Prey are animals hunted for food. Flying fish developed an odd way to escape **predators**. Predators are animals that hunt other animals for food.

Flying fish have fins. These fins are on each side. They're behind the **gills**. Gills are breathing **organs**. Organs are body parts.

Flying fish have extra-long fins. Their fins are as long as their bodies. They're used for jumping and gliding. It looks like they're flying.

Flying fish can glide farther than the length of a football field.

Cassowaries

Cassowaries are large birds that live in jungles in Australia. They look odd.

Their **crests** may help them run. Crests are feathers or tissue growths on top of a bird's head. A cassowary crest is like a helmet. Cassowaries stretch out their crests. Crests protect their heads. They avoid vines. They avoid predators.

The birds are about 6 feet (1.8 meters) tall. They can weigh up to 167 pounds (76 kg). They have strong, fast legs.

Cassowaries are fast runners.

Aye-Ayes

Aye-ayes are **primates**. Primates include humans, apes, and monkeys. They live in Madagascar near Africa.

Aye-ayes look odd. Their heads look like those of **rodents**. Rodents include rats and mice.

Aye-ayes are **nocturnal**. They hunt at night. They have skinny hands with long, bony middle fingers. Aye-ayes tap on wood. They listen for sounds. They bite holes in the wood. They use their middle fingers to dig out bugs.

Aye-ayes spend most of their lives in trees.

Platypuses

Platypuses are **mammals** that live in Australian rivers. Mammals have fur. Most give live birth to their young. But not platypuses! They lay eggs to produce young.

Platypuses also look different from most mammals. They have **bills**. Bills are beaks like ducks have. They have otter feet and bodies. They have beaver tails.

Platypuses are built to hunt underwater. They swim. They have special feelers on their bills. They can feel prey's movements.

Folds of skin cover platypuses' eyes and ears to prevent water from entering. Their nostrils are sealed closed.

Platypuses have double the fur! They have two layers of fur to stay warm and dry.

Males are venomous. They have sharp stingers on the back of their feet. Their venom can kill dogs. It causes great pain to humans. But they're only venomous during mating season.

Mammals usually aren't venomous. Platypuses are so odd. They're venomous mammals that lay eggs!

Humans Do What?!?

Michel Lotito was known as "Mr. Eats All." He ate 18 bikes. He ate 15 shopping carts. He ate 7 televisions. He ate 2 beds. He took 2 years to eat an airplane. He took these things apart. He cut them up. He drank mineral oil. This made things go down easier. He had pica. This is a medical condition. It causes people to crave dirt, glass, and metal. Pica can cause blockage, lead poisoning, and damage. But Lotito's stomach was odd. His stomach walls were double thick. His stomach juices were powerful. Sharp things could pass through his body.

Anglerfish

Anglerfish are bony fish. They live thousands of feet below the sea. They live in icy black water.

Females are predators. They go fishing for prey. Each female has a long, thin growth on its forehead. It looks like a fishing rod. It makes light. Anglerfish use it like a **lure**. They attract other fish to them. Then they eat them.

Male anglerfish attach themselves to the females.

Anglerfish skin reflects blue light. It helps camoflauge them from prey.

Female anglerfish have long teeth. Their teeth look like fangs. They have flexible stomachs. They can swallow prey twice their size.

Males are smaller than females. They have tiny teeth. Full grown males don't have any internal **organs**. Organs are body parts. Male anglerfish don't have rods. They attach themselves to the females.

Did You Know...?

- Some peoples of Madagascar were afraid of aye-ayes. They thought aye-ayes pointed their middle fingers at people who were going to die.

- Platypuses are bottom-feeders. They scoop up food along with bits of gravel and mud. They store this stuff in cheek pouches. They don't have teeth. Gravel helps break down their food.

Consider This!

Take a Position! Which animal do you think is the oddest? Why do you think so? Argue your point with reasons and evidence.

Think About It! All animals are odd. All animals are special. What does it really mean to be "odd"? What does it mean to be "normal"?

Learn More
- **Book:** Parker, Steve. 2016. Extreme Animals. Gareth Stevens Publishing LLLP.
- **Article:** Popular Mechanics - "The 40 Most Extreme Animals on the Planet" by Emily Shiffer. 2020: https://www.popularmechanics.com/science/animals/g28857063/most-extreme-animals/.

Glossary

adapt (uh-DAPT) change

bills (BILZ) bird beaks similar to those of ducks

crests (KRESTS) tufts of feathers or tissue growth on a bird's head

environment (ihn-VY-ruh-muhnt) home or surroundings

gills (GIHLZ) breathing organs of a fish

habits (HA-bihts) usual ways of doing things

lure (LOOR) something used to attract and catch something else

mammals (MA-muhls) animals with fur or hair that generally give live birth to their young

nocturnal (nok-TUR-nuhl) active at night

oddities (AH-duh-teez) things that look or appear strange

organs (OR-guhnz) body parts

predators (PREH-duh-turz) animals that hunt other animals for food

prey (PRAY) animals hunted for food

primates (PRY-mayts) animals including apes and monkeys

proboscis (pruh-BAH-suhs) a long, flexible nose

rodents (ROH-duhnts) small animals such as rats and mice

survive (sur-VYV) stay alive

Index

adaptations, 4, 5
anglerfish, 20–22
attacks, 9
aye-ayes, 14–15, 23

bills, 16
birds, 5, 12–13

cassowaries, 5, 12–13
crests, 12

defenses, 10, 12

eating, 8, 19

fins, 10–11
fish, 10–11, 20–22
flying fish, 10–11
frogs, 23

habitats, 4, 16, 20
habits, 4
humans, 19, 23
hunting, 14

Lotito, Michel, 19

mammals, 16–18
monkeys, 6–8

nocturnal animals, 14
noses, 6–8

pica, 19
platypuses, 16–18
poisonous animals, 18

predators and prey, 9, 10, 16, 20–22
primates, 6–8, 14–15, 23
proboscis monkeys, 6–8

running, 12–13

stomachs, 19, 22
swimming, 10–11, 16–18

wildcats, 9